S0-BBO-566

Who Wiggles This Tail?

by Cari Meister

PEBBLE
a capstone imprint

Whether light and fluffy or long and strong, animal tails are wonderful. Do you know who wiggles this tail? Guess, then turn the page to find out!

Who wiggles this tail?

BUNNY

A bunny's tail is small and fluffy.

Who wiggles this tail?

COW

A cow's tail swats away pesky flies.

Who wiggles this tail?

SKUNK

A bushy tail rises in warning. When threatened, the skunk will spray a smelly oil from its rear end. PEEE-U!

Who wiggles this tail?

ANGELFISH

A bright yellow tail swishes back and forth. It helps this fish glide through water.

Who wiggles this tail?

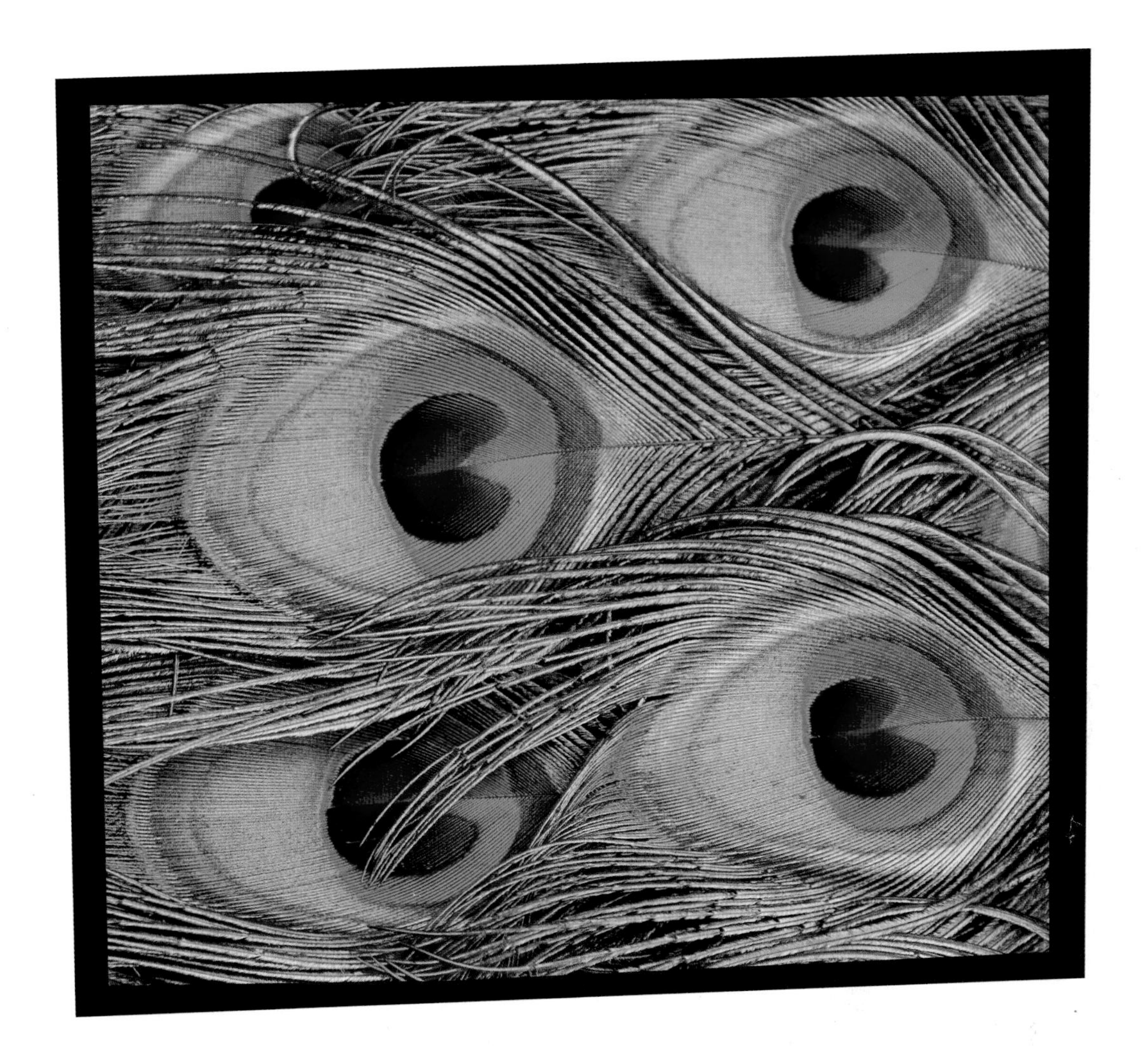

PEACOCK
This bird's beautiful tail
feathers charm the ladies
and attract a mate.

Who wiggles this tail?

RING-TAILED LEMUR

A lemur keeps its striped tail raised to tell others in its group to follow.

Who wiggles this tail?

CAT

A furry tail helps a cat balance on a tricky walk.

Who wiggles this tail?

HUMPBACK WHALE

This huge tail is called a fluke. The humpback slaps it on the water to communicate with other whales.

Who wiggles this tail?

RATTLESNAKE

Shake, shake, shake! A rattling tail warns animals to stay away from this dangerous snake.

Who wiggles this tail?

HIPPO

A hippo flings poop with its tail to mark its territory. YUCK!

Who wiggles this tail?

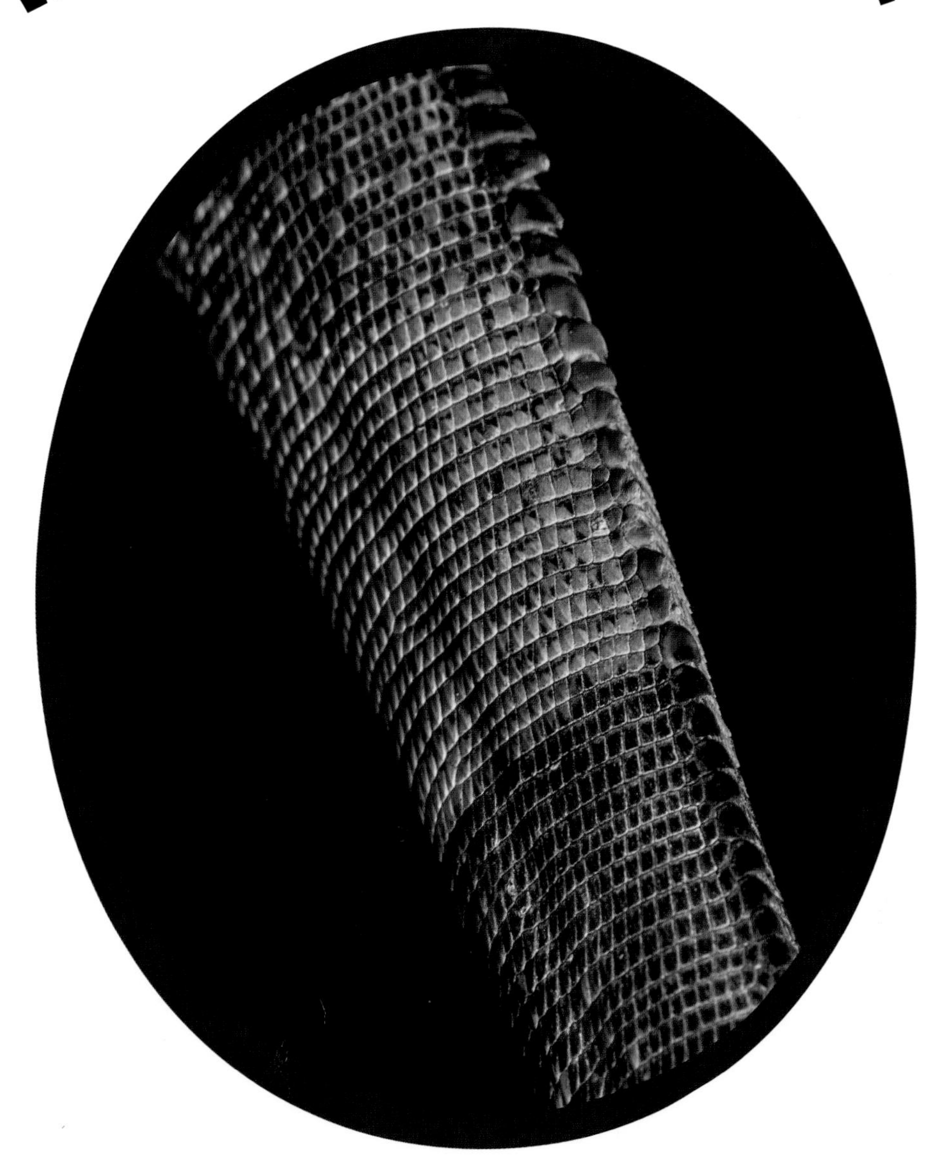

IGUANA

If this lizard loses its tail, the animal simply grows a new one.

Who wiggles this tail?

KANGAROO

A kangaroo can balance on its tail when boxing.

Who wiggles this tail?

FOX

A fox stays warm by wrapping its tail around parts of its body.

Who wiggles this tail?

A CHILD
That's SILLY! A child doesn't have a REAL tail. But pigtails are pretty close!

Tails are so cool!
They help animals
swim, balance, and
communicate.
Do you wish you
had a tail?

Pebble Sprout is published by Pebble, an imprint of Capstone.
1710 Roe Crest Drive
North Mankato, Minnesota 56003
www.capstonepub.com

Library of Congress Cataloging-in-Publication Data is available on the Library of Congress website.
ISBN 978-1-9771-2534-7 (library binding)
ISBN 978-1-9771-2544-6 (eBook PDF)

Summary: Swish! This photo-guessing game challenges pre-readers to guess whose tail appears in each image. The scaly, tufted, and bushy answers may be surprising!

Editor: Shelly Lyons
Designer: Bobbie Nuytten
Media Researcher: Jo Miller
Production Specialist: Katy LaVigne

Image Credits
Shutterstock: AKKHARAT JARUSILAWONG, 28, Arunee Rodloy, 10, Audrey Snider-Bell, 20, Breathes, 26, Charlotte Bleijenberg, Cover, cheapbooks, 29, Claude Huot, 22,Dai Mar Tamarack, 17, dejan_k, 5, Dennis Jacobsen, 25, Diane Kuhl, 6, Dicky Asmoro, 24, Dmitry Vorona, 15, ehtesham, 13, eshoot, 9, Fotos593, 23, Hajakely, 14, Heiko Kiera, 19, Holly Kuchera, 7, Ikoimages, 16, jgorzynik, 4, Marchenko Denis, 3, Miroslav Hlavko, 27, PeterVrabel, 12, Predrag Lukic, 11, Roman Samborskyi, 30, schankz, 21, sunsinger, 8, Wongymark1, 18

Design Elements
Capstone; Shutterstock: Artishok, cajoer, Fourleaflover, linear_design, srikorn thamniyom

Printed and bound in China.
3322

Good job! Try all the books in this series!